My Dad Loves Coffee

Written By: Daniel Hallback

My Dad Loves Coffee
Copyright © 2021 by Daniel Hallback
All Rights Reserved.

No part of this book may be used or reproduced by any means, graphic, electronic, or mechanical, including photocopying, recording, taping, or by any information storage retrieval system without the written permission of the publisher except in the case of brief quotations embodied in critical articles and reviews.

ISBN (Paperback) 978-1-955364-23-2
ISBN (Hardback) 978-1-955364-24-9
ISBN (Ebook) 978-1-955364-25-6

Vets Publish
www.vetspublish.com

When I wake up in the morning, my dad has a cup of coffee in his hand.

At lunch time my dad is drinking more coffee.

When we are on an airplane, the flight attendant brings him a cup of coffee.

At nighttime, he drinks a cup of decaf coffee.

He has a favorite coffee mug.

He has a favorite coffee shop.

He even eats coffee ice cream.

One day, my dad and I planted coffee trees together.

We water the coffee trees together.

After a couple of years, the coffee trees grew berries.

Together, we pick the berries from the trees.

Inside of the berries there are coffee beans.

We clean the coffee beans together.

We roast the beans together.

I love coffee and spending time with my dad.

The End